GW00600718

Australian
ANIMALS

Published by
Woollahra Sales and Imports
Unit 6, 32–60 Alice Street, Newtown, New South Wales, Australia, 2042
Phone: (02) 9557 8299 Facsimile: (02) 9557 8202
Email: wsi@pacific.net.au

Produced for the publisher by
Murray David Publishing
Publishing Director: Murray Child
Marketing Director: David Jenkins
35 Borgnis Street, Davidson, New South Wales, Australia, 2085

Text by Dalys Newman
Printed in Indonesia

ISBN 1 876553 33 2

Australian

ANIMALS

PHOTOGRAPHS BY KEN STEPNELL
TEXT BY DALYS NEWMAN

WOOLLAHRA

Introduction

Australia is unique among the continents for its strange assortment of animal life. The age of mammals had only just begun some fifty million years ago when the island continent was cut adrift from surrounding land masses. The few early groups of primitive mammals already established on the continent, forerunners of Australia's present marsupials and monotremes, found themselves unchallenged by the more advanced types of placental mammals that were appearing on the other continents. And so they thrived. Marsupials all but vanished from the rest of the world, but Australia remains the 'land of the living fossils', with marsupials, such as the koala, kangaroo and wombat inhabiting every ecological niche.

Other creatures of the wild that inhabit this vast country are just as fascinating and colourful as our pouched friends. There are over 700 species of birds in Australia, over half found nowhere else in the world. From the flightless emu, to dainty fairy penguins and raucous, brilliantly coloured parrots—the unusual habits, song and plumage of our birdlife are richly diverse.

About 160 distinct species of snakes are found in Australia, two-thirds of which are venomous, but fewer than twenty species really dangerous. There are about 180 species of frogs and two species of crocodiles—the largest reptiles found in the country. There are unique varieties of lizards amongst the geckos, skinks, dragons and goannas found in all corners of the country from the scorched desert plains to dense scrublands. Five species of marine turtles breed in Australian territory and there are at least thirteen native freshwater tortoises. Insects and spiders come in all shapes and sizes, from gigantic stick insects and bull-ants to bird-eating spiders and gloriously coloured rainforest butterflies.There are more than 1500 different species of spiders and over 360 species of butterflies.

Australia is truly blessed with her wildlife. Fossil evidence indicates that Australia's fauna has not changed to any great degree since remote times, but wider settlement of the continent is bringing increasing pressure to bear on wildlife, native mammals in particular . The conservation of the country's unique fauna has now become an important public issue with wildlife reserves being put aside and specialised habitats being conserved. The comparatively short list of extinct animals is being held to its present size and, with continued vigilance, the country's remarkable heritage of unique wildlife should stay intact.

Title page: The thorny devil (*Moloch horridus*) is found in the arid regions of Australia. **Above:** Spinifex hopping-mice *(Notomys alexis)* are a desert species found in central and western Australia. **Opposite:** One of Australia's best-loved attractions, the koala (*Phascolarctos cinereus*) is often seen sitting high in the branches of eucalypts which dominate the forests of the country's eastern seaboard.

Australia's most famous marsupial, the koala (*Phascolarctos cinereus*) is found in wet and dry eucalypt forests and woodlands along the eastern coast. Soft, grizzly-grey furry animals, they grow to about 75 centimetres long and are highly specialised for arboreal life. Their hands have a vice-like grip between the first two and other three fingers, and together with very long arms and curved claws, this enables them to climb smooth tree trunks. The tail is replaced by a callused pad which enables them to sit for long periods in tree forks without discomfort.

Gum leaves are the koalas sole diet and they consume about a kilogram of foliage each day. They are perhaps less nocturnal in habit than many other marsupials, waking up at any time during the day to start feeding on nearby leaves. Their slothful appearance is misleading as they can move rapidly through the branches and jump for a metre or so from branch to branch.

Koalas produce only one young at a time, the baby being only two centimetres long at birth with strong forelimbs to assist it on its journey to its mother's pouch. It does not appear outside the pouch until it is about six months old and about 18 centimetres long. It uses the pouch for another two months, afterwards being carried on its mother's back or hugged closely to her when resting or cold. Even when weaned to a diet of gum leaves the young koala stays with its mother until quite large.

Above and following pages: A forest dweller, the eastern grey kangaroo *(Macropus giganteus)* is found from south-eastern South Australia to Cape York. Their preferred habitat is dry sclerophyll forest, woodland and scrub-land with adjacent grassy areas.

Left: The red kangaroo *(Macropus rufus)*, probably the biggest living marsupial, is the most impressive of the great kangaroos. Found through out the plains and drier inland areas, it has the widest distribution of any kangaroo species. Although wide ranging, settlement and fencing have restricted their movements and drought proves a serious enemy. Gregarious in their habits, these kangaroos are usually found in small mobs of about a dozen or so, but may move in groups of up to two hundred. Mainly nocturnal, they spend their days under shady trees moving out into the open at night to find food and water.

Opposite: A female red kangaroo *(Macropus rufus)* and her joey. The lightly built female is a soft-toned smoky blue which has earned her the nickname.of blue flyer.

Right: The black-footed rock wallaby *(Petrogale lateralis)* is found in the semi-arid to arid rocky country of central and western Australia. They spend most of their day sleeping in sheltered areas amongst the rocks, but in cooler weather may emerge to bask in the sun.

Below: The largest and most strikingly coloured of the rock wallabies, the yellow-footed rock wallaby *(Petrogale xanthops)* is usually only found in the arid regions of South Australia and New South Wales. Favourite haunts are rugged rocky ranges with open woodland and acacia scrubland. This wallaby is distinguished from all others by the ring markings on its tail, bright yellow hind feet and forearms, and large furry yellow and white ears.

Above: The brush-tailed rock wallaby *(Petrogale penicillata)* can be seen on the cliffs and rocky slopes in the mountainous regions of New South Wales. This stoutly built wallaby is a rich dark colour and has a distinctive strongly brushed tail which is held out behind when he leaps over rocks, acting as a balancer. Another unique feature of the rock wallaby is its feet: it has large thick pads with rough granulations over the soles, similar to the tread on a sandshoe. An extremely agile creature, this wallaby sleeps by day in the shelter of rocks and feeds at night on native grasses and other vegetation.

Above: Whiptail wallabies *(Macropus parryi)* are found on grassy hillsides in wet and dry sclerophyll forests, from northern Queensland to north-eastern New South Wales. They are large slender animals with exceptionally long tapering black-tipped tails. Gregarious creatures, they move in groups of about fifty, grazing on native grasses, herbs and ferns during the early morning and late afternoon. They sleep for most of the day and at night in the shelter of a shrub or low tree.

Opposite: The small tammar wallaby *(Macropus eugenii)* has the distinction of being the first marsupial seen by European man in the Australian region. This was in 1629 on the Abrolhos Islands in Western Australia, when the Dutch ship *Batavia,* captained by Francisco Pelsaert, ran aground. He described them as 'large numbers of cats, which are creatures of miraculous form, as big as a hare'. Today, these wallabies are found in Western and South Australia and nearby islands. An adult tammar stands about 60 centimetres high and they travel almost silently through established runways or paths to graze in more open areas.

Left: Essentially a bush-loving and browsing animal, the red-necked wallaby *(Macropus rufogriseus)* haunts the brushes and heath country of the low coastal tablelands and the dense undergrowth of the ranges up to high altitudes. They are found on the eastern and south-eastern mainland and in Tasmania. Large and gracefully built, they vary in colouration but are distinguished from other wallabies by the contrast of the reddish nape and shoulders with the greyish-fawn colour of the back. They can grow to 1.8 metres in length and the males are noticeably larger than the females.

Right: Although at home in swampy or marshy country, the swamp wallaby *(Wallabia bicolor)* is also found on scrub-covered hillsides and mountain heights from tropical to cool-temperate climates. Large, stocky animals, they have a dark brown back, contrasting with a rich rusty yellowish belly and black tail.

Opposite: Commonly found in the sclerophyll forests and woodlands of Queensland, the black-striped wallaby (*Macropus dorsalis*) grows to about 2.4 metres long. They are usually seen in groups of about twenty, travelling along established tracks to grazing areas at night.

Below: The red-necked pademelon *(Thylogale thetis)* is found in rainforest and wet forest country in Queensland and New South Wales. Their habit of living in ferns and other thick vegetation, through which the animals make a maze of tunnels, means that seeing them in the wild is difficult. They have a crouching-hopping gait, well-suited to the thick undergrowth. When alarmed they thump out a warning signal with their hind feet—a habit common to most members of the kangaroo family.

Second in size only to the ostrich, the emu *(Dromaius novaehollandiae)* is found on the plains and open woodlands throughout the Australian mainland. These huge, flightless birds usually travel in small flocks, walking long distances to find food and are capable of speeds up to 50 kilometres per hour. At times of drought emus may invade outlying farms, causing considerable damage. Standing up to two metres tall, they have tiny wings and long, powerful armour-plated feet and legs enabling them to run over most obstacles with scarcely a falter in stride. The female lays up to twenty dark green eggs in a scant nest on the ground, leaving the male to incubate them and then care for the chicks.

The isolated highlands of Tasmania are the haunts of the ferocious looking Tasmanian devil *(Sarcophilus harrisii)*. This strange marsupial was one of the first mammals observed by the earliest settlers of Van Diemen's Land, and its forbidding appearance earned it the name of 'devil'. The devil's voice probably contributed to the satanic effect for the colonists—it is a whining growl followed by a snarling cough or a low yelling growl when angry. Related to native cats, this animal is about a metre long and has relatively short legs, a short broad muzzle, small eyes and broad rounded ears. Its thick fur is black or dark brown with white patches. Its favourite food is carrion and when this is lacking it feeds on other animals. Three or four young are born in the typically immature marsupial fashion and enter the rearward-facing pouch, where they remain for about fifteen weeks. Despite its appearance and manner, the Tasmanian devil is a shy animal and almost wholly nocturnal in habit.

The spiny anteater or short-beaked echidna *(Tachyglossus auleatus)* is, along with the platypus, an egg-laying monotreme, the most primitive of mammals. The echidna lays eggs yet suckles its young; it has no teats but exudes milk through its pores. The female grows a pouch only when required for carrying its young and after this function has been completed the pouch closes up. They have long tongues covered with a sticky substance, obtaining their food by thrusting this tongue into ant hills and withdrawing it covered with hundreds of ants. Echidnas are found throughout Australia and their natural habitat is open forests, scrublands and rocky areas—they are sometimes seen in suburban gardens.

Above and opposite: Lumbering and bear-like, the common wombat *(Vombatus ursinus)* is found along coastal forests of the south-eastern mainland and in Tasmania. Extremely compact animals, they can weigh up to 35 kilograms but rarely reach more than a metre in length. Enormously powerful forelimbs equipped with impressive claws enable wombats to create extensive burrows in which they shelter during the day, going forth at night to feed on grass, roots and herbage.

Left: The hairy-nosed wombat (*Lasiorhinus latifrons)* is distinguished from the common wombat by its distinctive hairy snout, bigger ears and fine silky fur. They are found on arid coastal and inland plains of Western and South Australia. This species adapts to its arid habitat by lowering its metabolic rate when food is sparse. They do not need to drink and can go without water for three or four months, sleeping in their deep, cool burrows, conserving their energy to avoid dehydration. Burrows are often clustered together to form extensive warrens, with entrances meeting in a large communal pit.

Above: Gould's monitor *(Varanus gouldii)*, also known as Gould's sand goanna, is found in sandy areas throughout most of Australia. It often hides in burrows in soft soil and it feeds on lizards and snakes.

Opposite: More commonly known as the tree goanna, the lace monitor *(Varanus varius)* is Australia's second largest lizard, growing to a length of two metres. (The perentie is the largest.) These creatures are arboreal and take to the trees when startled. Capable hunters, they feed on insects, other reptiles, small mammals and nesting birds. They also feed on carrion and several lace monitors may be seen feeding off the same carcass.

Right: At home on the water and on land, Merten's water monitor *(Varanus mertensi)* is found near the permanent waterways of central Queensland, the coastal Northern Territory and the Kimberley region of Western Australia. Most of its food—fish, frogs, crustaceans and large aquatic insects—is found beneath the water where it walks on the river bed with its eyes open.

Previous pages: Australia's largest and most powerful lizard, the perentie *(Varanus giganteus)* is found in the desert regions of northern and central Australia. It can grow up to 2.4 metres in length and feeds on carrion, snakes, small mammals and birds.

Above: The shingleback lizard *(Trachydosaurus rugosus)* is widely distributed in semi-deserts, savannah and open forests throughout the southern half of Australia. Its colour is variable, but it is usually a dull reddish brown, dark brown or black with scattered cream or yellow splotches. Grossly enlarged scales, like a pine cone in appearance, cover its tail and upper body. A slow-moving diurnal, it grows to 40 centimetres long and feeds on insects, snails, carrion, flowers, fruit and berries. Although fairly inoffensive, when cornered it adopts a U-shaped stance, aggressively flicks its tongue and may inflict a painful bite if sufficiently provoked.

Often encountered in suburban back gardens, the eastern blue-tongued lizard *(Tiliqua occipitalis)* is one of the largest skinks in the world. Stout and slow moving, it is a diurnal ground-dweller, feeding on a variety of insects, carrion, snails, wildflowers, native fruit and berries. It shelters at night in hollow logs and ground debris. If in danger the lizard usually retreats, but if cornered it will hiss violently, opening its mouth and producing its bright blue tongue in an attempt to frighten its aggressor. It is found in a wide variety of habitats in south-eastern South Australia, through Victoria, eastern New South Wales and most of Queensland to the northern Northern Territory and north-western Western Australia. The central blue tongue *(Tiliqua occipitalis multifasciata)*, pictured below, is found only in the arid regions of the country.

Above: Found living beside waterways along the eastern coast, the eastern water dragon *(Physignathus lesueurii)* is semi-aquatic and arboreal, often perching on tree branches overhanging the water in which it takes refuge if threatened. Male water dragons grow larger than females, specimens of 90 centimetres having been recorded. They feed on insects and aquatic organisms, including frogs, as well as other small terrestrial vertebrates, fruits and berries. Very competent swimmers, they propel themselves through the water by means of their strong tails.

Left: The semi-arboreal bearded dragon *(Amphibolurus barbatus)* is often seen on fence posts, basking in the sun. Camouflage is its principal form of protection but if molested it resorts to a spectacular defensive stance, opening its mouth fully to display the bright yellow interior, extending its 'beard' and expanding its ribs to create a greatly enlarged appearance. It grows to 75 centimetres and feeds on insects, worms, mice, small lizards and soft ground herbage. An egg layer, it deposits up to twenty-five eggs in a shallow hole scooped in sandy soil.

Above: Although fearsome in appearance, the thorny devil *(Moloch horridus)* is completely harmless, relying on its grotesque looks to ward off predators. When attacked it will retract its head under its body and present its enemies with the spiky hump on its neck. This slow-moving lizard inhabits the desert and semi-desert areas of central and south-western Australia. Well adapted to arid conditions, it can soak up water with its skin and feeds only on ants, often eating thousands during the course of one meal.

Above: The thick-tailed gecko *(Underwoodisaurus milii)* is usually found in the drier outback areas of Australia. Geckos are found in all parts of Australia except Tasmania, and they are all nocturnal. Their soft bodies, lacking the overlapping scales of most lizards, give them a fragile appearance. They are renowned for their habit of jettisoning their tails during stressful situations. The discarded, wriggling tail is intended to capture the attention of the aggressor while the gecko slithers away. Many geckos have adhesive discs on their toes which enable them to move effortlessly on smooth surfaces.

Above: The swift-moving painted dragon (*Amphibolurus pictus)* is a diurnal lizard found in the drier parts of the continent. During the mating season, the male is undoubtedly Australia's most colourful lizard, acquiring a bright blue flush on the throat and flanks and bright yellow or orange on the chest. Reaching about 25 centimetres in length, they occupy shallow burrows in sandy soils, usually at the base of saltbush or other low scrub, and feed on small insects.

Right: The peninsula dragon *(Ctenophorus fionni)* is one of about fifty species of dragon lizards found in Australia. All are egg-laying and, unlike snakes and geckos, they do not shed their tails in order to escape predators. Dragon lizards are characterised by round heads, stout bodies, powerful limbs, long tails and rough scales. The majority are terrestrial but a large number are skilled climbers and a few are semi-aquatic. Many are furnished with an impressive array of spines or spectacular adornments. They are diurnal creatures, partial to basking in the sun on top of termite mounds or fence posts.

Above: Found in the arid areas of Australia, the central carpet snake *(Morelia bredit)* is a harmless serpent of the Boa family which has no venom and kills by constricting its prey to death and then swallowing it. Feeding mainly on bandicoots and other marsupials, these snakes often hunt and shelter in trees, but may hunt on ground and shelter in the burrows of other animals. The female lays up to fifty eggs and then coils around them until they hatch.

Opposite: Found throughout Australia, the common death adder *(Acanthopis antarcticus)* is a secretive nocturnal snake who spends the day half-buried in sand, soil or litter, often at the base of trees or shrubs. One of the country's most deadly species, it possesses well-developed fangs and venom glands and strikes with amazing speed from a tense, flattened, coiled position. In the wild, death adders feed on native insectivorous reptiles, birds and mammals, which they capture by wiggling their tails with an insect-like movement to act as a lure. With an average length of 45 centimetres, these snakes have short stubby bodies with small, rat-like tails ending in a curved soft spine.

The banded rock python (*Liasis childreni childreni)* is one of Australia's smallest pythons, reaching a length of about 1.5 metres. Found in the top half of the country, they feed on mice, lizards and small birds. Pythons have no venom and kill their victims by first gripping them with their teeth and then coiling their body around them, slowly tightening until the breath is squeezed out of them and they are subdued enough to eat. Their jaws and head bones are so constructed that the snake can swallow an animal of improbable size.

Above: The most spectacularly adorned of Australia's reptiles, the frilled lizard (*Chlamydosaurus kingii*) uses its distinctive collar to ward off predators. When it is disturbed it will face the intruder and open its frill, at the same time opening its mouth to display the vivid yellow lining. The collar also serves as a food storage area. Any insects the lizard catches are stored in the folds of the collar until required.

Left: The tiger snake (*Notechis scutatus*) is one of Australia's most notorious venomous snakes. Up to 1.2 metres in length, the tiger normally hunts during the day but can sometimes be seen slithering along on warm nights. Its preferred food is frogs and tadpoles and it is often found along watercourses, swamps, lakes and wet mountain slopes.

Above: Averaging 1.2 metres in length, freshwater crocodiles (*Crocodylus johnstoni*) are smaller than the saltwater variety and are generally regarded as harmless. During the day they remain dormant, floating in the water or lying under foliage. At night they forage for fish, frogs, birds, crustaceans, small mammals and reptiles.

Right: Unique to Australia, the platypus (*Ornithorhynchus anatinus*) is an extraordinary mixture of reptile, bird and animal. An egg-laying mammal, this strange creature has a broad, flat snout, resembling a duck bill, a thick coat of brown fur, webbed feet with claws and a broad flat tail that aids in swimming. Found near riverbanks and lakes along the eastern seaboard and in Tasmania, they sleep most of the day in nests in burrows dug in riverbanks and feed at night on a variety of bottom-living invertebrates.

Above: Evolved from arboreal, possum-like ancestors, the Tasmanian bettong *(Bettongia gaimardi)* is abundant on this island where it inhabits dense grass cover in forest country. It sleeps by day in a well-constructed nest of grass and bark and forages at night for fungi, succulent roots and seeds. Solitary animals, they are intensely territorial and partly carnivorous.

Left: The fat-tailed dunnart *(Sminthopsis crassicaudata)* stores fat in its tail, taking advantage of a good season to eat as much as it can in preparation for the possible scarcity of food. They can also enter a state of torpor, similar to hibernation, when food is short, presumably a way of eking out the fat stored in their tails. These tiny creatures grow to about 10 centimetres and are found in arid to moderately wet woodland and shrubland over most of southern Australia.

Opposite: Relentless hunters of almost anything that moves, striped-faced dunnarts *(Sminthopsis macroura)* inhabit heaths and grasslands. These delicately built marsupials sleep during the day undercover or in soil cracks, and at night forage for insects and other arthropods. When threatened, they bare their efficient array of teeth and utter hissing sounds.

Above: Also known as the rabbit-eared bandicoot, the bilby *(Macrotis lagotis)* is one of the country's rarest marsupials, found mainly in shrubland and woodland in central Australia. The only bandicoot known to make burrows, they sleep by day in nests at the end of their long burrows and at night dig pits in the sand in search of burrowing insect larvae and succulent plant material. The female produces a litter of two who remain in the pouch for about eleven weeks and are weaned two weeks later. A full grown bilby is about 80 centimetres from head to tail and can weigh up to 2.5 kilograms.

Above: A desert species, spinifex hopping mice *(Notomys alexis)* are well-adapted to their environment and can exist without water. They avoid daytime heat, remaining in their complex burrows a metre or more down in the cool and usually damp earth. Emerging at night, they feed on seeds, roots, shoots and insects. Remarkably elongated hind feet distinguish this species from all other native rats and mice. They have evolved a two-footed leaping action, similar to that of kangaroos and their general appearance is that of a miniature kangaroo.

Right: Highly gregarious animals, plains mice *(Pseudomys australis)* construct large complex systems of colonial burrows, connected by runways on the surface. Found on the arid gibber plains of central and southwestern Australia they spend their days sleeping in burrows, with up to twenty animals inhabiting a single burrow system. When breeding, the group is reduced to a male and two or three females. They feed at night, mostly on seeds, and are well adapted to arid conditions as they do not need to drink.

Opposite: The long-nosed potoroo *(Potorous tridactylus)* is found in Tasmania and on the south-eastern mainland in wet sclerophyll forest, cool rainforest and heathland. By day, these solitary animals sleep in a nest of vegetation, coming out at night to dig in the soil for succulent roots, fungi and insect larvae. One of the fastest moving of all Australian mammals, potoroos move in an unusual fashion with their front legs tucked under their chests, their bodies horizontal and parallel to the ground and their rear legs driving them forward like a bullet.

Above: Once common in the arid and semi-arid areas of Australia, the rufous hare wallaby *(Lagorchestes hirsutus)* is now restricted to a sparse population in part of the Tanami Desert. They sleep by day in short burrows or deep excavations in the shelter of a hummock or low shrub and come out at night to graze on tough native grasses, shrubs and herbs.

Opposite: The southern brown bandicoot *(Isoodon obesulus)* is found in woodland and scrubland areas with low groundcover on the south-eastern mainland. Bandicoots, who are omnivores, have the front teeth of carnivores but the fused rear toes of herbivores—the latter being regarded as a climbing or prehensile adaptation of use in combing the fur which is much infested with lice and ticks. Fiercely territorial, these animals grow to 50 centimetres in length.

Above: Found only in a small area in Victoria and in Tasmania, the very agile eastern barred bandicoot *(Perameles gunni)* is on the endangered list of animals. Inhabiting semi-arid woodland, shrubland and dunes, it is the largest of the barred bandicoots, growing to about 40 centimetres, and is identified by its striped hindquarters and short, slender, mainly white tail.

Above: Once widespread on the mainland, the eastern quoll *(Dasyurus viverrinus)* now apparently survives only in Tasmania where it is fairly common. Almost wholly ground-dwellers, these native cats sleep through the day in small caves, rock crevices or hollow logs, emerging at dusk to hunt small birds, mice, rats, lizards, insects and young rabbits. They grow up to 70 centimetres in length and favour wet to dry sclerophyll forest habitats. The native cats differ from most other marsupials in lacking the well-developed pouch of the kangaroos, koalas, possums and bandicoots. The cat pouch is only a shallow furry depression containing the teats.

Left and opposite: The sturdy spotted-tailed quoll *(Dasyurus maculatus)* is the largest known marsupial carnivore on the mainland, with an adult measuring more than a metre from nose to tail tip. Ferocious and stubborn animals, they are stealthy, solitary creatures, with great climbing ability and remarkable hunting skills. Their food consists of any small animals, including rabbits and smaller mammals, birds and their eggs, and reptiles. Found throughout the eastern mainland and Tasmania, their preferred habitat is relatively wet forested areas.

A unique member of the wallaby family, the quokka *(Setonix brachyurus)* is found on Rottnest Island off Fremantle in Western Australia and, less plentifully, on Bald Island and the coastal south-western areas of the mainland. Apart from its small size, about that of a hare, the quokka is distinguished from all other pademelon wallabies by its short tail, short feet and short rounded ears which can hardly be seen above the long fur. It has a rather shaggy coat and its favourite haunts are swampy tracts and thickets. Dutch navigator Willem de Vlamingh named Rottnest Island after this animal, which he believed to be some species of rat; the word Rottnest meaning 'rat's nest'.

Above: Restricted to alpine habitats, the mountain pygmy possum *(Burramys parvus)* inhabits dense shrubs, usually under a canopy of snow gums. Small (25 centimetres) and vulnerable, these gregarious creatures sleep during the day in nests in bushes or crevices and feed at night on seeds, fruit and insects. In winter they often enter ski lodges for food and shelter and may become torpid for several days during periods of extreme cold or shortage of food. During warmer months they store seeds and fruits to be eaten later.

Opposite: The eastern pygmy possum *(Cercartetus nanus)*, although mouse-like in size, is a very different creature to a mouse with its prehensile tail and pouch to carry its young. Huge bulging eyes and sensitive ears guide them in their nocturnal hunt for insects, fruit, nectar and pollen which is gathered with a brush-tipped tongue. They sleep by day in woven nests in tree-holes and crevices and, as winter approaches, they store fat in their tails to tide them through the coldest months when they become torpid.

Following pages: The sugar glider *(Petaurus breviceps)* is found throughout the northern and eastern mainland and in Tasmania. The smallest of the gliders, it can glide for a remarkable distance of up to 50 metres. Gliders leap from tree to tree by means of their flight membranes, an extension of the body skin which is joined to the legs so that it is tightly stretched when the animal leaps from a tree with all four limbs outspread. They feed on insects, blossoms, native fruit, acacia gum and sap.

Above and right: Brush-tailed possums *(Trichosurus vulpecula)* are the most common of Australia's marsupials, ranging all over the mainland and in Tasmania. Pictured above and right, they are more adaptable than most pouched animals, living under many different conditions, sometimes taking up residence in a house roof. Males scent-mark territories and defend these against other males. Their thick, close, woolly fur is of economic value and they have suffered ruthless exploitation in the past. Fast, agile climbers, they are silvery grey in colour with a long bushy tail, long pointed ears and a rather pointed foxy face. At night they feed on eucalyptus leaves, fruit and blossoms.

Opposite: The Tasmanian brush-tail *(Trichosurus vulpecula fuliginosus)*, has a richer fur than the mainland possum.

A carnivore, the numbat *(Myrmecobius fasciatus)* feeds exclusively on termites, probing them out of their narrow galleries with its long ribbon-like sticky tongue which is flicked rapidly in and out of its long snout. It is an unusual animal, differing from other anteaters in its lack of powerful digging claws, and differing from all other marsupials in being entirely diurnal, sleeping at night in hollow logs. Relatively small and weak (less than 500 grams in weight and up to 50 centimetres in height), they do not have the strength to breach termite mounds made of cemented soil, instead they root around the soil surface, exposing the tunnels which the termites use to enter and leave the termitarium. A rare and endangered species, they are found in woodlands and savannah in Western Australia.

Above: Often seen feeding in paddocks or by the roadside, the sulphur-crested cockatoo *(Cacatua galerita)* is one of Australia's most familiar birds. Found in varied habitats, these birds often entertain themselves in the wild by aimlessly stripping leaves and bark from trees—the city-dwellers gaining a bad reputation for tearing up wooden furniture and verandahs and stripping putty from window frames. They are popular cage birds, due in part to their ability to mimic human speech.

Opposite: The red-tailed black cockatoo *(Calyptorhynchus banksii)* is found in the woodlands or trees near watercourses in the Northern Territory. They have extremely rowdy calls, and the young male takes four years to acquire his magnificent adult plumage of scarlet tail panels.

Above: A long upper bill distinguishes the long-billed corella *(Cacatua tenuirostris)* from the more common little corella. These noisy birds are seen in flocks of sometimes thousands in open timber country, grasslands and farms in south-eastern and south-western Australia. Ground feeders, they eat seeds from surface plants and dig for roots and corms. As a result, their heads and breasts are often dirt-stained.

Opposite: Relatively rare, the Major Mitchell or pink cockatoo *(Cacatua leadbeateri)* is found mostly in the arid or semi-arid interior of the country. A very beautiful pink-washed, white cockatoo, this bird has an upswept crest banded with scarlet and yellow. Often seen in the company of galahs, it has a stuttering falsetto cry and nests on decayed debris in tree hollows.

Following pages: Found near river margins and in flood areas in New South Wales and northern Victoria, the striking superb parrot (*Polytelis swainsonii*) is a ground-feeder as well as feeding on the seeds and blossoms of acacias and eucalypts. They lay four to six round white eggs in nests in deep hollow limbs of eucalypts near or over water.

Above: The mainly green male eclectus parrot *(Eclectus roratus)* is strikingly different from the female who has bright red and blue plumage. Conspicuous rainforest birds, found mainly on the eastern Cape York Peninsula, these parrots feed on the seeds, fruits and flowers of the forest canopy, returning at dusk to specific roosting trees. Their raucous screeching and squawking is a familiar rainforest sound. During courtship there is mutual preening and the male feeds the female. Nests are built in deep hollow trees and both birds attend to the chicks.

Opposite: Also known as the blood rosella, the king parrot *(Alisterus scapularis)* is found in coastal forests and river margins in eastern Australia. The male (pictured) is the only Australian parrot to have an unmarked bright scarlet head and body. The females are mostly dark green with red bellies. They travel in pairs or flocks, feeding in foliage and on the ground, and may become tame in farmyards and other places where food is available.

Left: Crimson rosellas *(Platycercus elegans)* are amongst the showiest of Australia's parrots. They nest in hollows, often very high in the tree, and lay five to eight rounded white eggs. Found in rainforests, woodlands, coastal scrub and fern gullies, they also frequent public parks and gardens where they are often quite tame. Their food includes seeds, soft fruits, nectar and some insects.

Opposite: Found on or near river systems and lakes in inland south-eastern Australia, the regent parrot *(Polytelis anthopeplus)* is a large, fast-flying bird. They inhabit river red gum forests, black box woodlands and adjacent cleared country, and are often seen gathering seeds from the yellow Hibbertia, a wildflower that grows in the Victorian mallee country.

Following pages: Little corellas *(Cacatua sanguinea)* are small, nearly crestless white cockatoos with short bills, pink stains between the bill and eye and a bare, blue-grey eye-patch. Travelling in pairs and small to immense flocks, they are often seen in trees by the water or noisily feeding on the ground. Found throughout much of the country, they frequent timbered water courses, dams and tanks, grasslands, sandhills and mangrove country. Nests are made in tree hollows and sometimes in cliff cavities or termite mounds.

Above: Nankeen kestrels *(Falco cenchroides)* are distinctive small falcons often seen hovering over paddocks and roadsides. Found in both country and city habitats throughout Australia, they soar around city buildings and frolic in air currents. They hover motionless, gradually drifting downwards, before suddenly dropping on prey. Eggs are laid on decayed debris in tree hollows, on bare earth in cliff cavities, on building ledges and occasionally in other bird's nests.

Opposite: The country's noisiest raptor, the brown falcon or cackling hawk *(Falco berigora)* utters a series of demented cacklings and screeches, particularly during display-flights when it tumbles and dives through the air. Widespread throughout Australia and Tasmania it is found in varied habitats and is often seen perching on poles and fence posts. It makes a sloping descent to seize its prey on the ground and often chases insects on foot. Lazy nest-builders, they usually use the nest of a crow or other hawk, laying two to four buff-white, reddish mottled eggs.

Australia's largest bird of prey, the wedge-tailed eagle *(Aquila audax)* has a wingspan of up to three metres. Paired for life, this eagle builds a large nest of branches and sticks, lined with fresh eucalyptus leaves, on a commanding situation, tree, rock or even on a high bush in the desert. They fly with easy, powerful wingbeats, gliding or soaring to great heights in majestic circles, with upswept wings. At rest, they are often seen on dead trees, telephone poles or on the ground. These birds prey chiefly on carrion, rabbits and other wildlife, seldom attacking livestock.

Above and right: Often mistaken for owls, tawny frogmouths *(Podargus strigoides)* are large nocturnal birds with very broad bills and large yellow eyes. Found throughout Australia, during the day they rest in absolute stillness, assuming stick-like immobility, on a tree limb or on the ground where they are well camouflaged. At night they hunt insects and small vertebrates by scooping them up in their enormous mouths. It makes a crude platform nest of twigs on a horizontal tree limb where three to four white eggs are laid. The sexes are paired for life.

Left: Australia's most common owl, barn owls *(Tyto alba)* hunt during the night, pouncing on mice, small birds, bats and insects. During the day they hide in tree hollows and rock caves and often nest in farm buildings where their rodent food can be found. Also known as the white owl, they have distinctive large facial discs and beautifully blended pure-white and brownish-gray plumage; when in flight they appear like pale ghosts.

Above: Widespread and familiar, the eastern yellow robin *(Eopsaltria australis)* builds well-camouflaged cup-shaped nests out of bark shreds, bound with spiders' web and lined with grass, thin twigs and leaves. They decorate the outside of the nest with strips of bark, lichen and moss. In this work of art are laid two to three apple-green eggs with brown and lilac splotches. When the young hatch, the nest and surrounding area are kept spotlessly clean.

Opposite: A beautiful and popular cagebird, the Gouldian finch *(Erythrura gouldiae)* can be either red-headed, black-headed or golden-headed. Found in northern and north-eastern Australia, they frequent grassy flats and trees near water, often gathering in the late afternoon to bathe or sit together. They feed in the grass and other low growth and are fond of hot sunshine, being active in the heat of the tropical day when most other birds are quiet.

Above: Commonly found in the northern coastal and inner coastal timber areas, the blue-winged kookaburra *(Dacelo leachii)* is renowned for stealing eggs and young from nearby nests and is constantly harassed by other birds. Also known as the barking or howling jackass, it has an appalling call, somewhat like a machine-driven hacksaw. Shyer than its more common relative, the laughing kookaburra, this bird nests in hollows high in trees where it lays three or four white eggs.

Opposite: The male superb blue wren *(Malurus cyaneus)* shows brilliant colour when in breeding plumage. One of the best known of the smaller bird families, these delightful little birds with their long cocked tails and happy trilling song are found throughout Australia, in suburban parks and gardens as well as in the undergrowth of open forests and woodlands. Wrens live in family groups and can produce many young when conditions are favourable. Sadly, their friendliness makes them vulnerable to attacks from domestic cats in suburban areas.

Opposite: Comical on land and graceful on water, the Australian pelican *(Pelecanus conspicillatus)* is found on large shallow waters, both coastal and inland, throughout the mainland and Tasmania. Pelicans waddle on land because their legs are set far apart and to the rear for efficient paddling; they fly and soar superbly, some diving for their prey and some swimming after it. Often seen perching on piles, logs and fishing boats, these birds are easily tamed when fed.

Below: Australia's only stork, the jabiru or black-necked stork *(Ephippiorhynchus asiaticus)* is a shy creature found in swamps and lagoons, mainly in northern areas. An impressive bird, standing over 1.25 metres, it travels over long distances, usually resting in swamps or marshes where it feeds on fish, frogs, crabs and carrion. They nest in live or dead trees, usually near water, where they lay two to four whitish eggs. If there are no high trees in the vicinity, it will build its nest on the ground.

Following pages: Large birds of the shallow, coastal waters, gannets *(Morus serrator)* are common around the southern Australian shores. Renowned for their diving skill, when fishing they make spectacular dives from up to 50 metres, submerging to about 9 metres and grasping their prey in their beaks from underneath as they rise. Sometimes a flock of these birds will round up fish into a concentrated shoal before diving for their meal.

The white-faced heron *(Egretta novaehollandiae)*, often incorrectly called the blue crane, is the heron familiar to most Australians. Very common almost wherever there is shallow water, these birds are often seen at roadsides, especially in wet years, sluggishly taking off when cars appear. They build untidy shallow nests in trees either over water or some distance away and lay three to five pale blue eggs. Their diet consists of yabbies, eels, frogs, small lizards and rodents.

Above: The Australian bustard (*Ardeotis australis*) is a ground-dwelling bird found in grasslands and pastoral country. Formerly widespread throughout most of the country it is now common only in areas away from settlement in parts of central, northern and western Australia. Lordly birds, they have a slow, stately walk, holding their head and bill tip up in the air. The male has a distinctive mating display, fanning his tail forward over his back, strutting around and uttering a hoarse, roaring call.

Opposite: Found in eucalypt woodlands and on the fringes of plains in south-eastern Australia and Tasmania, the musk lorikeet (*Glossopsitta concinna*) is a sturdily built bird with a short wedge-shaped tail. It has a loud penetrating call and prefers to feed in the larger blossoming gum trees, but will eat berries from a variety of native and introduced plants. In common with several other parrots, it exudes a musky odour, after which this species is named. It is seen in either pairs or flocks, in which the birds remain paired.

Opposite: A skilled fisherman, when hunting its prey the osprey (*Pandion haliaetus*) enters the water feet first with a spectacular splash. Found in coastal waters, its range has been drastically reduced because of toxic pesticides found in rivers and seas. Ospreys build extremely large nests of sticks and small branches, adding to them year after year. The nest is lined with seaweed and may be built on the ground, among rocks, or in trees near water.

Left: A bird of the tropics, the pied heron (*Ardea picata*) is found only in coastal swamps in the far north. These birds have long delicate breeding plumes, previously much sought after to decorate ladies' hats. Aggressive and belligerent, pied herons practise piracy, terrorising other birds and forcing them to disgorge food.

Below: The pied oystercatcher (*Haematopus longirostris*) is a large sturdy wading bird found in coastal areas throughout the country. They are regarded as shore birds rather than sea birds and are sometimes seen feeding on inland fields. Their straight, stout builds are used to pry shellfish off rocks, open oysters and dismember crustaceans.

Above: Named after the sound of its call, the wompoo pigeon (*Ptilinopus magnificus*) is a magnificent large bird found in the rainforests of coastal eastern Australia. It eats wild figs and other fruits and builds very flimsy nests on palm fronds or in leafy twigs, often over a stream.

Left: A distinctive slow soaring hawk, the brahminy kite (*Haliastur indus*) is found along the coastlines of northern Australia, especially near sandflats and mangroves. Often quite tame, these scavengers seek carrion along the tide-lines, occasionally catching fish, reptiles and insects. It has a loud whining call and nests high up in swamp trees.

Above: A wading bird with long slender pink legs, the pied stilt (*Himantopus himantopus*) is found in swamps, estuaries, mudflats and shallows throughout Australia. With legs measuring 20 centimetres, this is the longest legged bird, proportionate to its body size, in the country.

Above: Found throughout Australia in coastal and inland shallows, the white or large egret *(Ardea alba)* is by far the tallest egret in the country. A gaunt lanky bird, it is usually solitary but is sometimes seen in small parties and during the wet season in the far north very large companies of egrets inhabit the floodplains.

Opposite: One of only two cranes found in the country, the brolga *(Grus rubicunda)* is widespread in northern, eastern, south-eastern and inland Australia. These birds are renowned for their group dancing displays accompanied by whooping trumpet calls. Inhabiting shallow swamps, floodplains, grasslands and ploughed fields, they plunge their heads under water to dig for roots and corms.

Following pages: Huge flocks of black swans *(Cygnus atratus)* can be found on large open waters, either fresh or salt, throughout Australia. These highly nomadic birds swim with their necks arched and often carry their feathers or wings raised in an aggressive display. They fly strongly, with whistling wings and baying, trumpeting calls. After breeding, they moult and become temporarily flightless. Black swans are often found on ornamental lakes where they breed freely and frequently become tame. They construct their nests out of reeds, grasses and weed in shallow water or on islands and lay four to seven greenish white eggs.

Three species of seal breed on the mainland and offshore islands of Australia: the sea lion, Australian fur seal and New Zealand fur seal.
Below: Australian fur seal pups *(Arctocephalus pusillis)*.
Right and Opposite: The only member of the seal family unique to Australia, the sea lion *(Neophoca cinerea)* is found in the cool, temperate South Australian waters and off the coast of south-western Australia. When on land they occupy sandy beaches and form breeding colonies in rocky areas; over five hundred inhabiting a colony at Seal Bay on Kangaroo Island in South Australia. They were virtually wiped out by sealers but have recently returned to the island where they are now protected. The males, who can reach 2.5 metres, are much larger than the females.

Above: The female loggerhead turtle *(Caretta caretta)* comes ashore to lay about fifty eggs in a shallow nest pit. She then covers the eggs and heads back to sea, using her flippers in a typically four-legged animal fashion. Named for their big heads, which can be up to 30 centimetres long, these turtles have large brown eyes and pronounced beaks. Loggerheads are almost entirely carnivorous, feeding on crabs, molluscs, sponges, jelly fish and sometimes algae. They are found in the tropical and warm temperate waters off the coast, including the Great Barrier Reef. There is a unique turtle rookery at Mon Repos on the coast near Bundaberg in Queensland, where over a thousand loggerheads come from November to February to lay their eggs.

Above: One of the most formidable creatures in the country, the saltwater crocodile *(Crocodylus porosus)* is found in northern seas, estuaries, rivers and pools. This large (up to six metres) and active predator is a very aggressive hunter, feeding on fish, waterfowl and animals such as kangaroos and wallabies. After making their kill they stash their victim under water, lodged in snags and river banks, returning to feed when the carcass begins to decompose. On occasion they become man-eaters. These crocodiles hunt mainly at night and can often be seen during the day basking on riverbanks.

Right: The eastern long-necked tortoise *(Chelodina longicollis)* is found in fresh water from Cape York to Victoria. Shy, unobtrusive animals, they spend much of their time submerged in rivers and pools but will travel long distances in search of ephemeral swamps, especially during summer rains. Although aquatic in habit, they have powerful legs and can cover the ground with speed and determination.

Above: Found in all eastern states, sometimes in large numbers, the spotted marsh frog *(Limnodynastes tasmaniensis)* is the most abundant and widespread of Australia's land frogs. Growing to about four centimetres, this small frog is attractively coloured with large olive-green spots on a lighter green background. It breeds in flooded paddocks, permanent waters and culverts and the female produces her spawn in a mass of foam which carries the eggs.

Opposite: White-lipped tree frogs *(Litoria infrafrenta)* are found in moist cool places across the northern half of Australia, often in association with human habitation. Active climbers, they have a suction pad or disc on the end of each toe which assists in climbing. This frog spends his days in a moist cranny, emerging to hunt small creatures at night. When the rain falls the males start calling, a deep, repeated 'crawk', which is an important courtship function attracting other males and the females to a certain breeding area. Females lay 200 to 2000 eggs after rainfalls in the summer months and the mottled brown or green tadpoles take about six weeks to mature to frogs.

Above: Burrowing frogs *(Cyclorana mainii)* are well-adapted to life in Australia's arid interior. They burrow into the drying mud of inland water holes and re-appear months or years later when rains again fill the inland watercourses.

Opposite: Found in Victoria, Tasmania, New South Wales and South Australia, the green and gold bell frog *(Litoria raniformis)* is mostly aquatic. They are usually seen in vegetation in or beside permanent water, such as swamp reedbeds or paperbark trees around lakes or lagoons. The males usually call while floating in open water. Australia has green and gold bell frogs in both south-eastern and south-western regions, some species being nearly identical.

Following pages: Often seen in breeding groups in the creeks and streams of south-eastern Australia and Tasmania, the whistling tree frog *(Litoria ewingi)* is one of the few species of frog to call throughout the year. Its call is a rapid, pulsing whistle, hence its name. Found in a variety of habitats from swamps, lagoons, wet and dry sclerophyll forest to alpine grasslands and bogs, these frogs grow up to 4 centimetres in size and are variable in colour pattern.

The only penguin to breed in Australia, the little or fairy penguin *(Eudyptula minor)* is a popular tourist attraction at its breeding colony on Phillip Island, in Victoria. Every evening, during a particular stage in their breeding cycle, they waddle ashore to the burrows they vacated in the morning for their fishing expeditions at sea. Very much smaller than their relatives in the Antarctic, fairy penguins stand about 25 centimetres high. Breeding burrows are from 45 to 90 centimetres long and about the size of a rabbit burrow. Here, they lay two or three large white eggs and the male and female take turns sitting on them while the other swims all day in search of food which includes small marine life and tiny fish. Powerful swimmers, penguins literally 'fly under water' propelled by their stiff, flattened flippers and using their webbed feet and small tails as rudders. They have very thick feathers which are an effective insulation against saltwater.

Opposite and above: There are about fifty species of stick insects (order *Phasmatodea*) in Australia. The goliath phasmid found in Queensland, is one of the world's largest insects, measuring up to 22 centimetres in length. Stick insects, with their powerful cutting jaws, can be devastating consumers of green foliage, at times causing such heavy defoliation that the affected trees die. Masters of camouflage, they look like sticks and it is often impossible to see them amongst the branchlets.

Overleaf: The ministral mantid *(Orthodera ministralis)* is one of the many species of mantids found in Australia, most of them inhabiting the subtropics and tropics. All mantids are predatory and eat other insects, the female of the species often eating the males at mating time. They have sickle-shaped forelegs with toothed edges, well adapted for seizing prey, and their large compound eyes protrude beyond the sides of their head, enabling binocular vision.

Above: A large, robust arachnid, the garden or orb-weaving spider *(Eriophora biapicata)* constructs perfect orb-webs between bushes and trees and across pathways and doors. The spider is usually to be found hiding nearby or waiting on the web. Conspicuous only during the summer months, they are not dangerous spiders—bites have been recorded but have not been serious.

Left: One of Australia's biggest spiders, the barking or whistling spider *(Selenocosmia crassipes)* is known to attack and eat frogs and birds, including chickens and young quail. They make a clearly audible barking sound when disturbed or irritated. When making this sound, they usually rear up on their hind legs like a funnelweb spider, to which they are related. Despite their aggressive appearance, they are not dangerous to humans. These spiders spend the day in an underground burrow up to 60 centimetres deep, and are found throughout the inland regions and hilly eastern Australian areas.

Above: The silk-lined burrow of the trap-door spider (*Arbantis* spp.) is inhabited by the female, the male only being admitted during mating. Many species in this family build solid doors to their burrows out of silk secretions, the door is then hinged to the ground by more silk.

Above and opposite: Belonging to the order of Odonata, 300 of the 4000 known species of dragonflies are found in Australia. Large, robust insects, they are powerful fliers with long thin wings which are held stiffly out from their bodies and large eyes which all but encompass their heads. Known to reach speeds of up to 80 kilometres an hour, they catch and eat insects on the wing, cradling them in their legs which are placed well forward on the thorax. Primitive aquatic insects, they breed in freshwater habitats, the female usually laying her eggs directly into the water. The underwater larval stage, known as a mud-eye, feeds on other water insects. When ready to emerge, the fully fed nymph crawls above the water onto a tree trunk or reed and the adult emerges in similar fashion to the cicada. The adult then flies away from the home pool.

Following pages: There over 200 species of termites in Australia, some of which build conspicuous mounds, often as high as 6 metres. The mounds are built of earth particles joined together with saliva and, when dry, are as hard as concrete. A maze of channels, rooms and passages make up the interior. The mounds give the termites some control over their environment, particularly during prolonged dry spells.

Index